KU-216-457

OFFICIAL SQA PAST PAPERS WITH ANSWERS

STANDARD GRADE | GENERAL

MATHEMATICS
2006-2010

2006 EXAM – page 3
General Level 2006 Paper 1
General Level 2006 Paper 2

2007 EXAM – page 33
General Level 2007 Paper 1
General Level 2007 Paper 2

2008 EXAM – page 63
General Level 2008 Paper 1
General Level 2008 Paper 2

2009 EXAM – page 93
General Level 2009 Paper 1
General Level 2009 Paper 2

2010 EXAM – page 127
General Level 2010 Paper 1
General Level 2010 Paper 2

© Scottish Qualifications Authority
All rights reserved. Copying prohibited. No part of this publication may be reproduced, stored in a retrieval system, or transmitted in any form or by any means, electronic, mechanical, photocopying, recording or otherwise.

First exam published in 2006.
Published by Bright Red Publishing Ltd, 6 Stafford Street, Edinburgh EH3 7AU
tel: 0131 220 5804 fax: 0131 220 6710 info@brightredpublishing.co.uk www.brightredpublishing.co.uk

ISBN 978-1-84948-099-4

A CIP Catalogue record for this book is available from the British Library.

Bright Red Publishing is grateful to the copyright holders, as credited on the final page of the book, for permission to use their material.
Every effort has been made to trace the copyright holders and to obtain their permission for the use of copyright material.
Bright Red Publishing will be happy to receive information allowing us to rectify any error or omission in future editions.

[BLANK PAGE]

FOR OFFICIAL USE

G

KU	RE

Total
marks

2500/403

NATIONAL
QUALIFICATIONS
2006

FRIDAY, 5 MAY
10.40 AM – 11.15 AM

MATHEMATICS
STANDARD GRADE
General Level
Paper 1
Non-calculator

Fill in these boxes and read what is printed below.

Full name of centre

Town

Forename(s)

Surname

Date of birth
Day Month Year

Scottish candidate number

Number of seat

1 You may **not** use a calculator.

2 Answer as many questions as you can.

3 Write your working and answers in the spaces provided. Additional space is provided at the end of this question-answer book for use if required. If you use this space, write clearly the number of the question involved.

4 Full credit will be given only where the solution contains appropriate working.

5 Before leaving the examination room you must give this book to the invigilator. If you do not you may lose all the marks for this paper.

SCOTTISH
QUALIFICATIONS
AUTHORITY

©

FORMULAE LIST

Circumference of a circle: $C = \pi d$

Area of a circle: $A = \pi r^2$

Curved surface area of a cylinder: $A = 2\pi r h$

Volume of a cylinder: $V = \pi r^2 h$

Volume of a triangular prism: $V = Ah$

Theorem of Pythagoras:

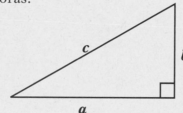

$$a^2 + b^2 = c^2$$

Trigonometric ratios
in a right angled
triangle:

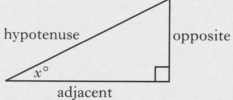

$$\tan x° = \frac{\text{opposite}}{\text{adjacent}}$$

$$\sin x° = \frac{\text{opposite}}{\text{hypotenuse}}$$

$$\cos x° = \frac{\text{adjacent}}{\text{hypotenuse}}$$

Gradient:

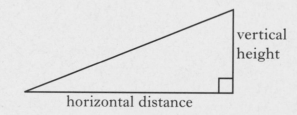

$$\text{Gradient} = \frac{\text{vertical height}}{\text{horizontal distance}}$$

Marks	KU	RE

1. Carry out the following calculations.

(a) $2 \cdot 73 + 7 \cdot 6 - 8 \cdot 4$

1		

(b) 13×7000

1		

(c) $56 \cdot 5 \div 500$

1		

(d) 30% of 92 litres

2		

[Turn over

DO NOT
WRITE IN
THIS
MARGIN

Marks | KU | RE

2.

Paulo's Pizzas

Student Discount

$\frac{1}{3}$ *off the price of each pizza*

Emily is a student and she buys a pizza from Paulo's Pizzas.

She chooses a pizza which is normally £8·49.

How much will Emily pay for the pizza?

3

3. A new movie costs $320 million to make.

Write this amount in scientific notation.

2

4. Jenni is making a wallpaper border.

She is using stars and dots to make the border.

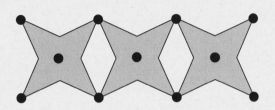

(a) Complete the table below.

Number of stars (s)	1	2	3	4	5
Number of dots (d)			11		

(b) Write down a formula for calculating the number of dots (d), when you know the number of stars (s).

(c) Each star is 10 centimetres long.

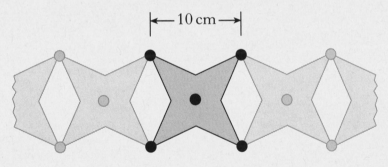

The wallpaper border Jenni makes is 300 centimetres long.

　(i) How many stars does Jenni need?

　(ii) How many dots does she need?

Marks KU RE

5. The line AB is drawn on the grid below.

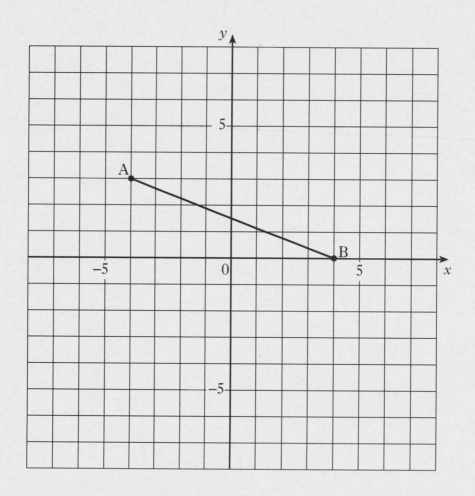

Calculate the gradient of the line AB.   2

6. A box contains 10 coloured balls.

There are 4 yellow balls, 3 blue balls, 2 green balls and 1 red ball.

(*a*) David takes a ball from the box.

What is the probability that the ball is blue?

1

(*b*) The ball is put back in the box.

2 yellow balls and the red ball are then removed.

What is the probability that the next ball David takes from the box is green?

2

[Turn over

DO NOT WRITE IN THIS MARGIN

Marks KU RE

7. The temperature in a supermarket freezer during a 12-hour period is shown in the graph below.

Temperature of Supermarket Freezer

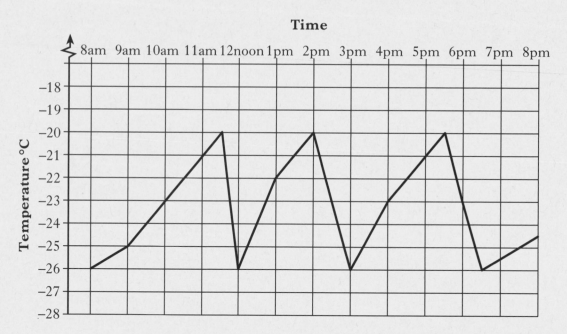

(a) From 8am, how long did it take for the temperature to rise to −20 °C?

1

(b) For how long, in **total**, was the temperature rising during the 12-hour period?

3

Marks | KU | RE

8. Rachel asks 19 friends how many text messages they sent last week.

Their answers are shown below.

34	25	46	62	28
38	42	23	25	15
32	52	35	44	30
10	33	41	55	

(*a*) Display Rachel's friends' answers in an ordered stem and leaf diagram.

3

(*b*) What is the median number of text messages?

1

[Turn over for Question 9 on *Page ten*

DO NOT
WRITE IN
THIS
MARGIN

Marks | KU | RE

9.

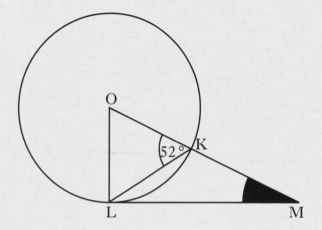

In the diagram above with circle centre O:

- LM is a tangent to the circle at L
- OM intersects the circle at K
- Angle OKL = 52°.

Calculate the size of the shaded angle OML.

3

[END OF QUESTION PAPER]

ADDITIONAL SPACE FOR ANSWERS

ADDITIONAL SPACE FOR ANSWERS

FOR OFFICIAL USE

G

	KU	RE
Total marks		

2500/404

NATIONAL
QUALIFICATIONS
2006

FRIDAY, 5 MAY
11.35 AM – 12.30 PM

MATHEMATICS
STANDARD GRADE
General Level
Paper 2

Fill in these boxes and read what is printed below.

Full name of centre

Town

Forename(s)

Surname

Date of birth
Day Month Year   Scottish candidate number   Number of seat

1 **You may use a calculator.**

2 Answer as many questions as you can.

3 Write your working and answers in the spaces provided. Additional space is provided at the end of this question-answer book for use if required. If you use this space, write clearly the number of the question involved.

4 Full credit will be given only where the solution contains appropriate working.

5 Before leaving the examination room you must give this book to the invigilator. If you do not you may lose all the marks for this paper.

SCOTTISH
QUALIFICATIONS
AUTHORITY

©

FORMULAE LIST

Circumference of a circle: $C = \pi d$

Area of a circle: $A = \pi r^2$

Curved surface area of a cylinder: $A = 2\pi rh$

Volume of a cylinder: $V = \pi r^2 h$

Volume of a triangular prism: $V = Ah$

Theorem of Pythagoras:

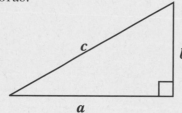

$$a^2 + b^2 = c^2$$

Trigonometric ratios
in a right angled
triangle:

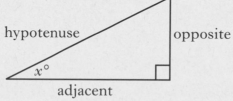

$$\tan x^\circ = \frac{\text{opposite}}{\text{adjacent}}$$

$$\sin x^\circ = \frac{\text{opposite}}{\text{hypotenuse}}$$

$$\cos x^\circ = \frac{\text{adjacent}}{\text{hypotenuse}}$$

Gradient:

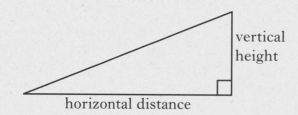

$$\textbf{Gradient} = \frac{\textbf{vertical height}}{\textbf{horizontal distance}}$$

Marks | KU | RE

1. The Sharkey family is going on holiday to France.

They will stay at the "Prenez Les Bains" campsite.

Prenez Les Bains	Tent holiday		Mobile Home holiday	
Start Date	Cost for 7 nights	Cost per extra night	Cost for 7 nights	Cost per extra night
26 June – 2 July	495	39	585	58
3 July – 9 July	535	41	615	65
10 July – 30 July	645	46	825	72
31 July – 13 Aug	699	47	880	75
14 Aug – 28 Aug	670	39	845	73

The family chooses a mobile home holiday.

Their holiday will start on 15 July and the family will stay for 12 nights.

Use the table above to calculate the cost of the holiday.

3

[Turn over

DO NOT WRITE IN THIS MARGIN

Marks | KU | RE

2. Carly bought a new printer for her computer.

The time taken to print a document is proportional to the number of pages printed.

It takes 7 minutes to print a document with 63 pages.

How many pages can be printed in half an hour?

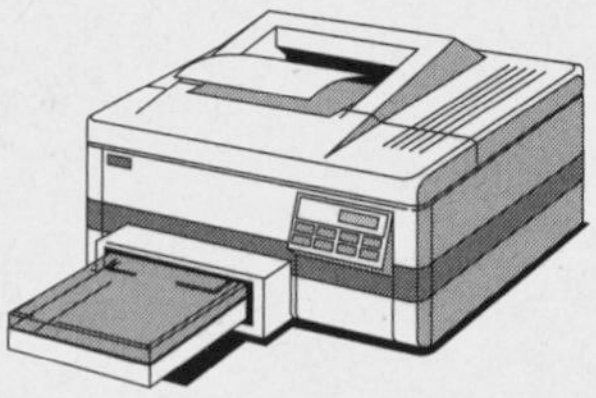

3

DO NOT WRITE IN THIS MARGIN

3. At a school fun day, prizes can be won by throwing darts at a target.

Each person throws **six** darts.

Points are awarded as follows.

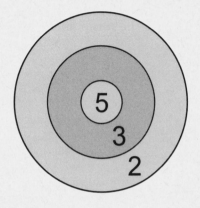

	POINTS
Centre	5
Middle Ring	3
Outer Ring	2
Miss	0

Prizes are won for **25 points or more**.

Complete the table below to show all the different ways to win a prize.

Number of darts scoring 5 points	Number of darts scoring 3 points	Number of darts scoring 2 points	Number of darts scoring 0 points	Total Points
4	2	0	0	26

Marks: 4

[Turn over

Marks | KU | RE

4. The entrance to a building is by a ramp as shown in the diagram below.

The length of the ramp is 180 centimetres.

The angle between the ramp and the ground is 12°.

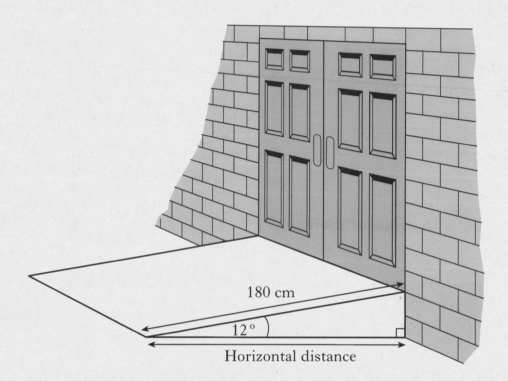

180 cm

12°

Horizontal distance

Calculate the horizontal distance.

Round your answer to one decimal place.

Do not use a scale drawing.

4

Marks   KU   RE

5. Ann works in a hotel.

She is paid £5·60 per hour on weekdays and double time at weekends.

Last month her gross pay was £436·80.

Ann worked a total of 54 hours on weekdays.

How many hours did she work at double time?

4

[Turn over

DO NOT WRITE IN THIS MARGIN

Marks   KU   RE

6. (*a*) Factorise

$$6a + 15b.$$

2

(*b*) Solve algebraically

$$4x - 3 = x + 21.$$

3

Marks | KU | RE

7. Amy and Brian travel from Dundee to Stonehaven.

The distance between Dundee and Stonehaven is 80 kilometres.

Amy takes 1 hour 30 minutes to travel by car.

Brian takes the train which travels at an average speed of 60 kilometres per hour.

What is the difference between their journey times?

4

[Turn over

Marks KU RE

8. ABCD is a rhombus.

AE = 4·3 metres and BE = 2·9 metres.

Calculate the perimeter of the rhombus.

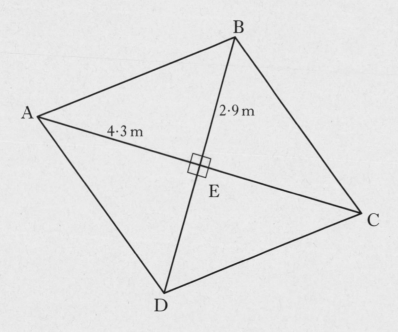

Do not use a scale drawing.

4

Marks | KU | RE

9. The top of Calum's desk is in the shape of a quarter-circle as shown.

The measurement shown is in metres.

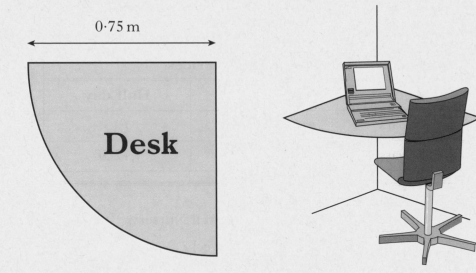

0·75 m

Desk

(*a*) Calculate the area of the top of the desk.

2

(*b*) Calum wants to paint the top of his desk.

The tin of paint he buys has a coverage of $1\,m^2$.

Using this tin of paint, how many times could he paint the top of his desk?

2

10. Maria is two years old.

Each week she goes to the nursery for 3 full days and 2 half days.

(a)

Playwell Nursery

Age	Prices	
	Full day	**Half day**
0–2 years	£28	£15
3–5 years	£23·50	£12·50

Maria's mother pays for her to attend Playwell Nursery.

How much does Maria's mother pay each week?

2

On Monday, Tuesday and Wednesday Maria goes to nursery from 9 am to 3 pm.

On Thursday and Friday she goes from 9 am to 12 noon.

(b) The nursery introduces a new hourly rate.

> **New Rate   £5 per hour**

Will Maria's mother save money when the nursery changes to the hourly rate?

Give a reason for your answer.

3

DO NOT
WRITE IN
THIS
MARGIN

Marks | KU | RE

11. The diagram below shows the positions of Lossiemouth and Leuchars.

A ship in the North Sea is on a bearing of 110° from Lossiemouth and 075° from Leuchars.

Show the position of the ship on the diagram below.

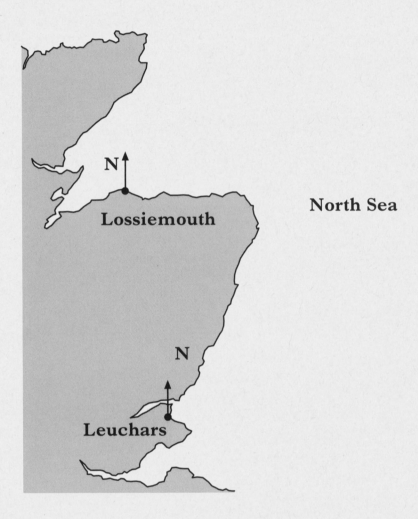

3

[Turn over for Question 12 on *Page fourteen*

Marks | KU | RE

12. Gordon is insuring his car with Carins Insurance.

The basic annual premium is £765.

As Gordon is a new customer his premium is calculated by taking $\frac{1}{5}$ off the basic annual premium.

However, because he wants to pay in monthly instalments, Carins Insurance add an extra 8% to his premium.

How much in total will Gordon pay per month?

4

[END OF QUESTION PAPER]

ADDITIONAL SPACE FOR ANSWERS

ADDITIONAL SPACE FOR ANSWERS

STANDARD GRADE | GENERAL

2007

[BLANK PAGE]

FOR OFFICIAL USE

G

	KU	RE
Total marks		

2500/403

NATIONAL
QUALIFICATIONS
2007

THURSDAY, 3 MAY
10.40 AM – 11.15 AM

MATHEMATICS
STANDARD GRADE
General Level
Paper 1
Non-calculator

Fill in these boxes and read what is printed below.

Full name of centre

Town

Forename(s)

Surname

Date of birth

| Day | Month | Year | | Scottish candidate number | | Number of seat |

1 **You may not use a calculator.**

2 Answer as many questions as you can.

3 Write your working and answers in the spaces provided. Additional space is provided at
  the end of this question-answer book for use if required. If you use this space, write clearly
  the number of the question involved.

4 Full credit will be given only where the solution contains appropriate working.

5 Before leaving the examination room you must give this book to the invigilator. If you do
  not you may lose all the marks for this paper.

SCOTTISH
QUALIFICATIONS
AUTHORITY

FORMULAE LIST

Circumference of a circle: $C = \pi d$

Area of a circle: $A = \pi r^2$

Curved surface area of a cylinder: $A = 2\pi rh$

Volume of a cylinder: $V = \pi r^2 h$

Volume of a triangular prism: $V = Ah$

Theorem of Pythagoras:

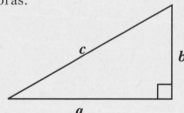

$$a^2 + b^2 = c^2$$

Trigonometric ratios
in a right angled
triangle:

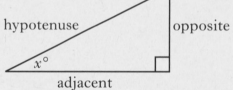

$$\tan x^\circ = \frac{\text{opposite}}{\text{adjacent}}$$

$$\sin x^\circ = \frac{\text{opposite}}{\text{hypotenuse}}$$

$$\cos x^\circ = \frac{\text{adjacent}}{\text{hypotenuse}}$$

Gradient:

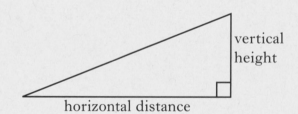

$$\text{Gradient} = \frac{\text{vertical height}}{\text{horizontal distance}}$$

DO NOT WRITE IN THIS MARGIN

	Marks	KU	RE
1. Carry out the following calculations.			
(*a*) $4\cdot27 - 1\cdot832$	1		
(*b*) $6\cdot53 \times 40$	1		
(*c*) $372 \div 8$	1		
(*d*) $5 \times 4\frac{1}{3}$	2		
2. A particle is radioactive for $2\cdot3 \times 10^{-4}$ seconds. Write this number in full.	2		

Marks KU RE

3. Zoe is a member of a gym.

The gym offers the following exercise sessions.

Exercise	Session Time
Weights	15 minutes
Dance	40 minutes
Running	20 minutes
Cycling	30 minutes
Swimming	45 minutes

Zoe is advised to choose **three** different exercises.

She wants to exercise for a **minimum of 90 minutes**.

One possible combination of three different exercises is shown in the table below.

Complete the table to show all the possible combinations of three different exercises Zoe can choose.

Weights	Dance	Running	Cycling	Swimming	Total Time (minutes)
		✓	✓	✓	95 minutes

3

DO NOT
WRITE IN
THIS
MARGIN

Marks   KU   RE

4. Complete this shape so that it has quarter-turn symmetry about O.

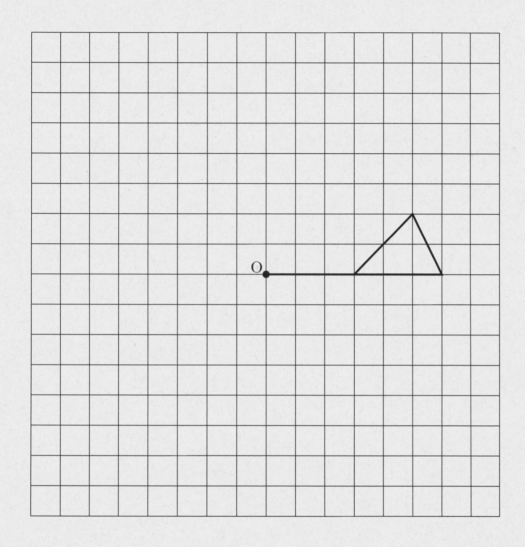

3

[Turn over

Marks | KU | RE

5. In an experiment Rashid measures the temperature of two liquids.

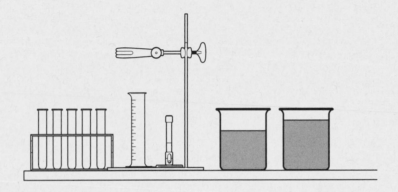

The temperature of the first liquid is −11° Celsius.

The temperature of the second liquid is 23° Celsius.

Find the difference between these temperatures.

2

Marks | KU | RE

6. A children's play area is to be fenced.

The fence is made in sections using lengths of wood, as shown below.

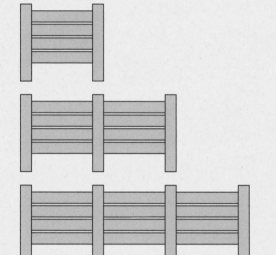

1 section

2 sections

3 sections

(a) Complete the table below.

Number of sections (s)	1	2	3	4	5		12
Number of lengths of wood (w)	6	11					

2

(b) Write down a formula for calculating the number of lengths of wood (w), when you know the number of sections (s).

2

(c) A fence has been made from 81 lengths of wood.

How many sections are in this fence?

You must show your working.

2

DO NOT WRITE IN THIS MARGIN

Marks | KU | RE

7. The table below shows the marks scored by pupils in French and Italian exams.

Pupil	A	B	C	D	E	F	G	H
French Mark	15	23	50	38	40	42	70	82
Italian Mark	28	31	62	54	45	55	85	95

(*a*) Using these marks, draw a scattergraph.

2

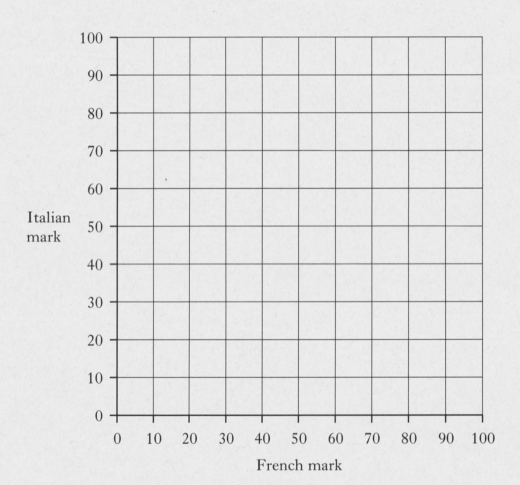

(*b*) Draw a best-fitting line on the graph.

1

Marks

KU RE

7. (continued)

(*c*) A pupil who scored 65 in his French exam was absent from the Italian exam.

Use your best-fitting line to estimate this pupil's Italian mark.

1

8. Pamela sees a bracelet costing £65 in a jeweller's window.

The jeweller offers Pamela a 5% discount.

Pamela decides to buy the bracelet.

How much does she pay?

3

[Turn over

DO NOT WRITE IN THIS MARGIN

Marks | KU | RE

9. Craig works in the school office.

Shown below is his order for 25 boxes of folders.

Office Supplies	
Blue Folders	7 boxes
Green Folders	11 boxes
Pink Folders	3 boxes
Yellow Folders	4 boxes
Total	**25 boxes**

His order has arrived in identical boxes but they are not labelled.

(*a*) What is the probability that the first box Craig opens contains pink folders? **1**

(*b*) The first box Craig opens contains green folders.

What is the probability that the next box he opens contains blue folders? **2**

Marks | KU | RE

10. There are 720 pupils in Laggan High School.

The ratio of boys to girls in the school is 5 : 4.

How many girls are in the school?

3

[END OF QUESTION PAPER]

ADDITIONAL SPACE FOR ANSWERS

FOR OFFICIAL USE

G

	KU	RE
Total marks		

2500/404

NATIONAL
QUALIFICATIONS
2007

THURSDAY, 3 MAY
11.35 AM – 12.30 PM

**MATHEMATICS
STANDARD GRADE**
General Level
Paper 2

Fill in these boxes and read what is printed below.

Full name of centre

Town

Forename(s)

Surname

Date of birth
Day Month Year   Scottish candidate number   Number of seat

1 **You may use a calculator.**

2 Answer as many questions as you can.

3 Write your working and answers in the spaces provided. Additional space is provided at the end of this question-answer book for use if required. If you use this space, write clearly the number of the question involved.

4 Full credit will be given only where the solution contains appropriate working.

5 Before leaving the examination room you must give this book to the invigilator. If you do not you may lose all the marks for this paper.

SCOTTISH
QUALIFICATIONS
AUTHORITY

©

FORMULAE LIST

Circumference of a circle: $C = \pi d$

Area of a circle: $A = \pi r^2$

Curved surface area of a cylinder: $A = 2\pi rh$

Volume of a cylinder: $V = \pi r^2 h$

Volume of a triangular prism: $V = Ah$

Theorem of Pythagoras:

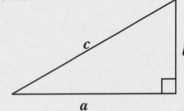

$$a^2 + b^2 = c^2$$

Trigonometric ratios
in a right angled
triangle:

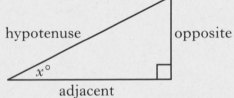

$$\tan x^\circ = \frac{\text{opposite}}{\text{adjacent}}$$

$$\sin x^\circ = \frac{\text{opposite}}{\text{hypotenuse}}$$

$$\cos x^\circ = \frac{\text{adjacent}}{\text{hypotenuse}}$$

Gradient:

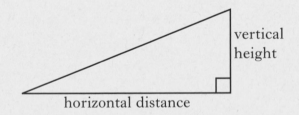

$$\text{Gradient} = \frac{\text{vertical height}}{\text{horizontal distance}}$$

Marks | KU | RE

1. A Sprinter train travels at an average speed of 144 kilometres per hour.

The train takes 1 hour 15 minutes to travel between Dingwall and Aberdeen.

Calculate the distance between Dingwall and Aberdeen.

2

[Turn over

DO NOT WRITE IN THIS MARGIN

Marks | KU | RE

2. Mr McGill is a bricklayer.

He builds a wall using 7500 bricks:

- each brick costs 23 pence
- a charge of £200 is made for every 500 bricks he lays.

What is the **total** cost of building the wall? 3

DO NOT WRITE IN THIS MARGIN

Marks | KU | RE

3.

BELMONT VETS
CHECK-UP FEES

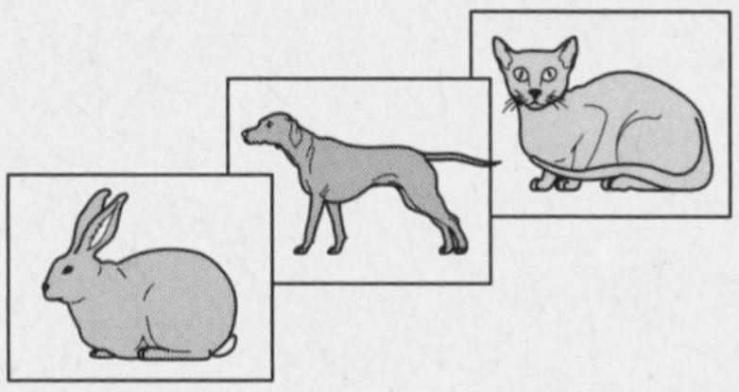

Dog	**£17·50**
Cat	**£11·75**
Rabbit	**£7·95**

The Wilson family owns two dogs and a cat.

Last year each dog had two check-ups at Belmont Vets.

The family cat also had check-ups last year.

The Wilson's total check-up fees for the two dogs and the cat were £105·25.

How often did the cat have a check-up?

4

[Turn over

DO NOT
WRITE IN
THIS
MARGIN

Marks | KU | RE

4. A rectangular metal grill for a window is shown below.

Two diagonal metal bars strengthen the grill.

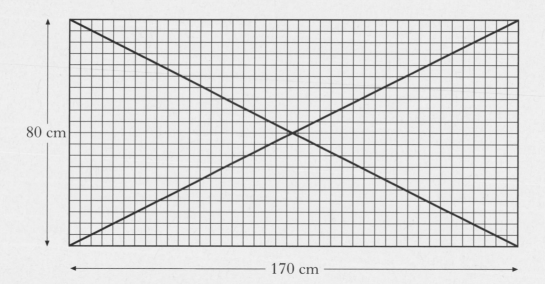

Find the length of one of the metal bars.

Round your answer to the nearest centimetre.

Do not use a scale drawing.

4

DO NOT WRITE IN THIS MARGIN

Marks KU RE

5. (*a*) Simplify

$$2(3x + 7) + 4(3 - x).$$

3

(*b*) Solve the inequality

$$4a - 3 \geq 21.$$

2

[Turn over

Marks | KU | RE

6. DEFG is a kite:

- angle DEG = 35°
- EF = 14 centimetres.

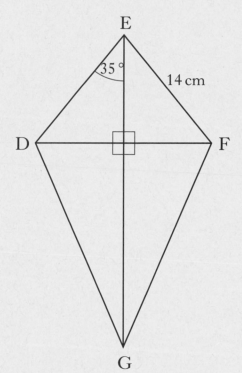

Calculate the length of DF.

4

Marks KU RE

7. A supermarket has a canopy over its entrance.

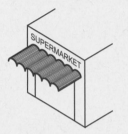

The edge of the canopy has 6 semicircles as shown below.

← 4 m →

Each semicircle has a diameter of 4 metres.

(*a*) Find the length of the curved edge of **one of the semicircles**.

2

(*b*) Tony attaches fairy lights to the edge of the canopy.

He has 40 metres of fairy lights.

Is this enough for the whole canopy?

Give a reason for your answer.

2

DO NOT
WRITE IN
THIS
MARGIN

Marks KU RE

8.

**Platinum
Saver Account**

6·3% interest per annum

Sally invests £4200 in the Platinum Saver Account which pays 6·3% interest per annum.

How much simple interest will she receive after 10 months?

3

Marks | KU | RE

9. In the diagram:

- O is the centre of the circle
- AC is a diameter
- B is a point on the circumference
- angle BAC = 43°.

Calculate the size of shaded angle BOC.

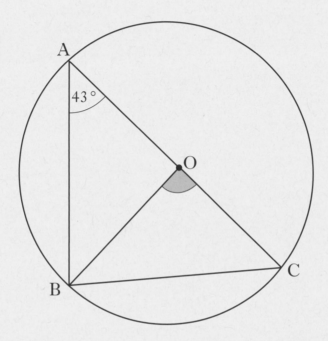

3

Marks | KU | RE

10. The end face of a grain hopper is shown in the diagram.

   (*a*) Calculate the area of the end face.

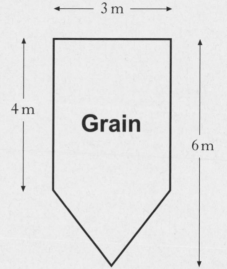

   (*b*) The grain hopper is in the shape of a prism with a length of 3·5 metres.

   Find the volume of the hopper.

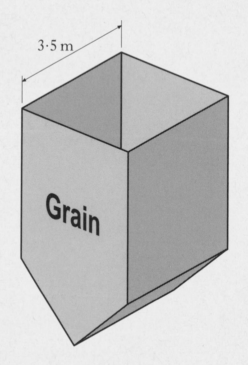

3

2

DO NOT
WRITE IN
THIS
MARGIN

Marks | KU | RE

11. The diagram below shows the design for a house window.

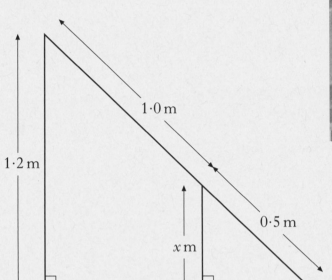

Find the value of x.

3

[Turn over for Question 12 on *Page fourteen*

DO NOT WRITE IN THIS MARGIN

Marks	KU	RE

12. The burning time, t minutes, of a candle varies directly as its height, h millimetres.

A candle with a height of 75 millimetres burns for 180 minutes.

(*a*) What is the burning time of a 40 millimetre candle? **3**

(*b*) A candle burns for $2\frac{1}{2}$ hours.

What is the height of this candle? **3**

[*END OF QUESTION PAPER*]

ADDITIONAL SPACE FOR ANSWERS

ADDITIONAL SPACE FOR ANSWERS

STANDARD GRADE | GENERAL

2008

[BLANK PAGE]

FOR OFFICIAL USE

G

	KU	RE
Total marks		

2500/403

NATIONAL
QUALIFICATIONS
2008

THURSDAY, 8 MAY
10.40 AM – 11.15 AM

MATHEMATICS
STANDARD GRADE
General Level
Paper 1
Non-calculator

Fill in these boxes and read what is printed below.

Full name of centre

Town

Forename(s)

Surname

Date of birth

Day Month Year   Scottish candidate number   Number of seat

1 **You may not use a calculator.**

2 Answer as many questions as you can.

3 Write your working and answers in the spaces provided. Additional space is provided at the end of this question-answer book for use if required. If you use this space, write clearly the number of the question involved.

4 Full credit will be given only where the solution contains appropriate working.

5 Before leaving the examination room you must give this book to the invigilator. If you do not you may lose all the marks for this paper.

FORMULAE LIST

Circumference of a circle: $\qquad$ $C = \pi d$

Area of a circle: $\qquad$ $A = \pi r^2$

Curved surface area of a cylinder: $\qquad$ $A = 2\pi rh$

Volume of a cylinder: $\qquad$ $V = \pi r^2 h$

Volume of a triangular prism: $\qquad$ $V = Ah$

Theorem of Pythagoras:

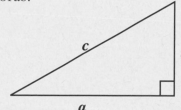
$$a^2 + b^2 = c^2$$

Trigonometric ratios
in a right angled
triangle:

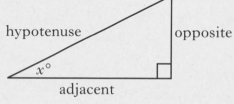

$$\tan x^\circ = \frac{\text{opposite}}{\text{adjacent}}$$

$$\sin x^\circ = \frac{\text{opposite}}{\text{hypotenuse}}$$

$$\cos x^\circ = \frac{\text{adjacent}}{\text{hypotenuse}}$$

Gradient:

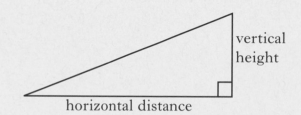

$$\textbf{Gradient} = \frac{\textbf{vertical height}}{\textbf{horizontal distance}}$$

DO NOT
WRITE IN
THIS
MARGIN

Marks | KU | RE

1. Carry out the following calculations.

(*a*) $12 \cdot 76 - 3 \cdot 18 + 4 \cdot 59$

1

(*b*) $6 \cdot 39 \times 9$

1

(*c*) $8 \cdot 74 \div 200$

1

(*d*) $\dfrac{5}{6}$ of 420

2

[Turn over

Marks | KU | RE

2. In the "Fame Show", the percentage of telephone votes cast for each act is shown below.

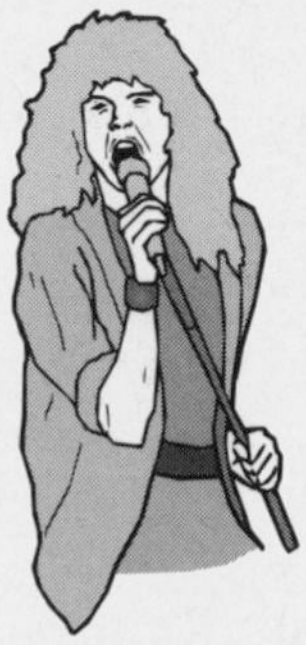

Plastik Money	23%
Brian Martins	35%
Starshine	30%
Carrie Gordon	12%

Altogether 15 000 000 votes were cast.

How many votes did Starshine receive?

3

DO NOT
WRITE IN
THIS
MARGIN

Marks | KU | RE

3. AB and BC are two sides of a kite ABCD.

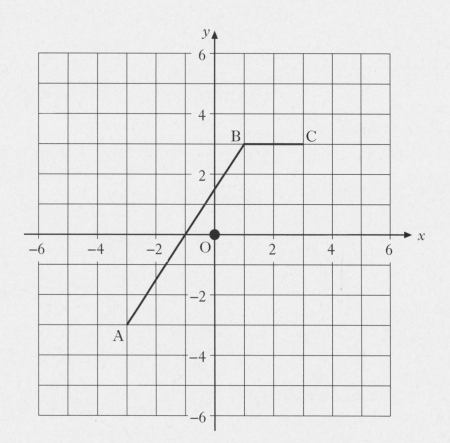

(*a*) Plot point D to complete kite ABCD.

1

(*b*) Reflect kite ABCD in the **y-axis**.

3

Marks | KU | RE

4. Europe is the world's second smallest continent.

Its area is approximately 10 400 000 square kilometres.

Write this number in scientific notation.

2

5. Samantha is playing the computer game "Castle Challenge".

To enter the castle she needs the correct four digit code.

The computer gives her some clues:

- only digits 1 to 9 can be used
- each digit is greater than the one before
- the sum of all four digits is 14.

(*a*) The first code Samantha found was 1, 3, 4, 6.

Use the clues to list all the possible codes in the table below.

1	3	4	6

3

(*b*) The computer gives Samantha another clue.

- three of the digits in the code are prime numbers

What is the four digit code Samantha needs to enter the castle?

1

[Turn over

DO NOT
WRITE IN
THIS
MARGIN

Marks | KU | RE

6.

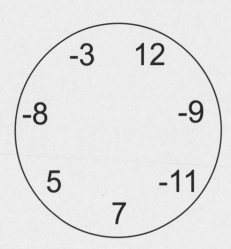

The circle above contains seven numbers.

Find the three numbers from the circle which add up to −10.

You must show your working.

3

DO NOT WRITE IN THIS MARGIN

Marks | KU | RE

7. The cost of sending a letter depends on the size of the letter and the weight of the letter.

Format	Weight	Cost	
		1st Class Mail	**2nd Class Mail**
Letter	0–100 g	34p	24p
Large Letter	0–100 g	48p	40p
	101–250 g	70p	60p
	251–500 g	98p	83p
	501–750 g	142p	120p

Claire sends a letter weighing 50 g by 2nd class mail.

She also sends a large letter weighing 375 g by 1st class mail.

Use the table above to calculate the total cost.

3

[Turn over

DO NOT
WRITE IN
THIS
MARGIN

Marks KU RE

8. Four girls and two boys decide to organise a tennis tournament for themselves.

Each name is written on a plastic token and put in a bag.

(*a*) What is the probability that the first token drawn from the bag has a girl's name on it?

1

(*b*) The first token drawn from the bag has a girl's name on it.

This token is **not** returned to the bag.

What is the probability that the next token drawn from the bag has a boy's name on it?

2

Marks | KU | RE

9.

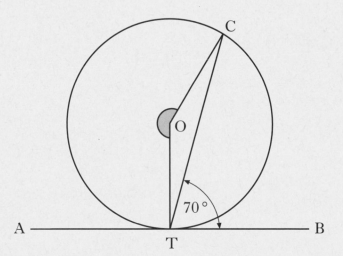

In the diagram above:

- O is the centre of the circle
- AB is a tangent to the circle at T
- angle BTC = 70°.

Calculate the size of the shaded angle TOC.

3

[END OF QUESTION PAPER]

ADDITIONAL SPACE FOR ANSWERS

FOR OFFICIAL USE

G

Total marks

KU	RE

2500/404

NATIONAL QUALIFICATIONS 2008

THURSDAY, 8 MAY 11.35 AM – 12.30 PM

MATHEMATICS
STANDARD GRADE
General Level
Paper 2

Fill in these boxes and read what is printed below.

Full name of centre

Town

Forename(s)

Surname

Date of birth

Day Month Year   Scottish candidate number   Number of seat

1 **You may use a calculator.**

2 Answer as many questions as you can.

3 Write your working and answers in the spaces provided. Additional space is provided at the end of this question-answer book for use if required. If you use this space, write clearly the number of the question involved.

4 Full credit will be given only where the solution contains appropriate working.

5 Before leaving the examination room you must give this book to the invigilator. If you do not you may lose all the marks for this paper.

FORMULAE LIST

Circumference of a circle: $C = \pi d$

Area of a circle: $A = \pi r^2$

Curved surface area of a cylinder: $A = 2\pi rh$

Volume of a cylinder: $V = \pi r^2 h$

Volume of a triangular prism: $V = Ah$

Theorem of Pythagoras:

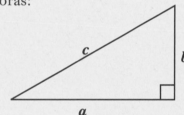

$$a^2 + b^2 = c^2$$

Trigonometric ratios
in a right angled
triangle:

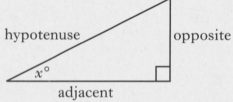

$$\tan x^\circ = \frac{\text{opposite}}{\text{adjacent}}$$

$$\sin x^\circ = \frac{\text{opposite}}{\text{hypotenuse}}$$

$$\cos x^\circ = \frac{\text{adjacent}}{\text{hypotenuse}}$$

Gradient:

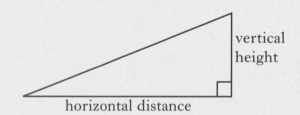

$$\text{Gradient} = \frac{\text{vertical height}}{\text{horizontal distance}}$$

DO NOT WRITE IN THIS MARGIN

Marks | KU | RE

1. Corrina has a part time job in a local pottery.

She paints designs on coffee mugs.

Her basic rate of pay is £6·25 per hour.

She also gets paid an extra 22 pence for every mug she paints.

Last week Corrina worked 15 hours and painted 40 mugs.

How much was she paid?

3

[Turn over

DO NOT
WRITE IN
THIS
MARGIN

Marks | KU | RE

2. Charlie's new car has an on-board computer.

At the end of a journey the car's computer displays the information below.

Journey information

distance **157.5 miles**

average speed **45 miles/hour**

Use the information above to calculate the time he has taken for his journey.

Give your answer in hours and minutes.

4

Marks   KU   RE

3.

Ben needs 550 grams of flour to bake two small loaves of bread.

(*a*) How many **kilograms** of flour will he need for thirteen small loaves?

2

Ben buys his flour in 1·5 kilogram bags.

(*b*) How many bags of flour will he need to bake the thirteen small loaves?

1

[Turn over

Marks | KU | RE

4. Mhairi makes necklaces in M-shapes using silver bars.

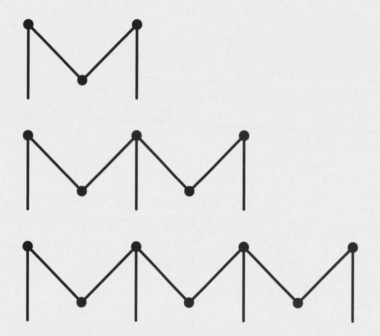

(*a*) Complete the table below.

Number of M-shapes (*m*)	1	2	3	4		15
Number of bars (*b*)	4	7				

2

(*b*) Write down a formula for calculating the number of bars (*b*) when you know the number of M-shapes (*m*).

2

(*c*) Mhairi has 76 silver bars.

How many M-shapes can she make?

2

Marks | KU | RE

5. Lewis is designing a bird box for his garden.

The dimensions for the side of the box are shown in the diagram below.

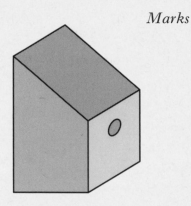

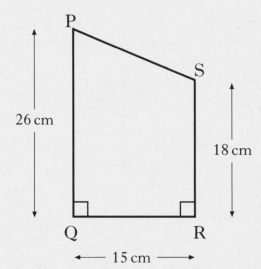

Calculate the length of side PS.

Do not use a scale drawing.

4

[Turn over

DO NOT WRITE IN THIS MARGIN

Marks	KU	RE

6. Gordon buys an antique teapot for £95.

He sells it on an Internet auction site for £133.

Calculate his percentage profit.

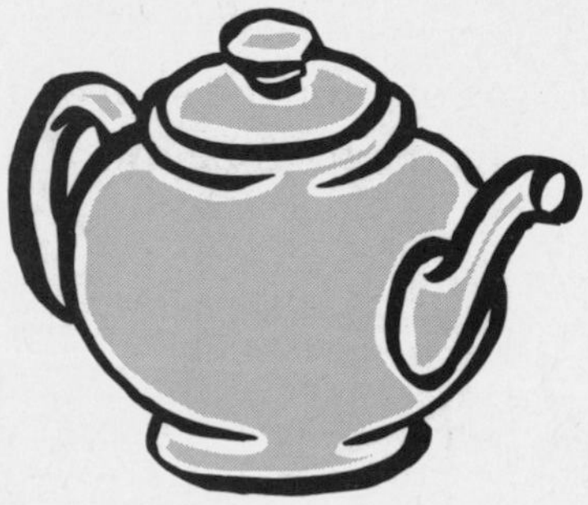

3

Marks | KU | RE

7. A piece of glass from a stained glass window is shown below.

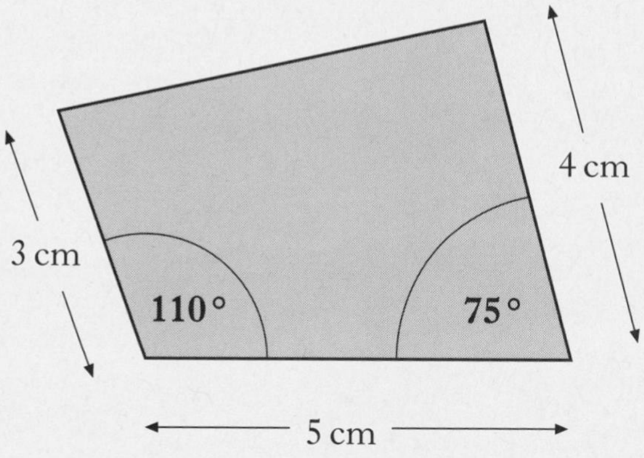

A larger piece of glass, the same shape, is to be made using a scale of 2:1.

Make an accurate drawing of the larger piece of glass.

3

[Turn over

DO NOT
WRITE IN
THIS
MARGIN

Marks | KU | RE

8. (*a*) Solve algebraically

$$7t - 3 = t + 45.$$

3

(*b*) Factorise fully

$$20x - 12y.$$

2

Marks | KU | RE

9. Ian is making a sign for Capaldi's Ice Cream Parlour.

The sign will have two equal straight edges and a semi-circular edge.

Each straight edge is 2·25 metres long and the radius of the semi-circle is 0·9 metres.

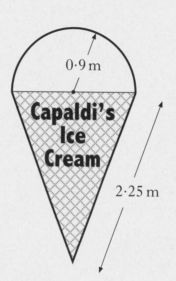

0·9 m

2·25 m

Calculate the perimeter of the sign.

4

[Turn over

DO NOT WRITE IN THIS MARGIN

Marks KU RE

10. Natalie wanted to know the average number of hours cars were parked in a car park.

She did a survey of 100 cars which were parked in the car park on a particular day.

Her results are shown below.

Parking time (hours)	*Frequency*	*Parking time* × *frequency*
1	28	
2	22	
3	10	
4	15	
5	11	
6	5	
7	9	
	Total = 100	Total =

Complete the above table and find the mean parking time per car.

3

Marks | KU | RE

11. Circular tops for yoghurt cartons are cut from a strip of metal foil as shown below.

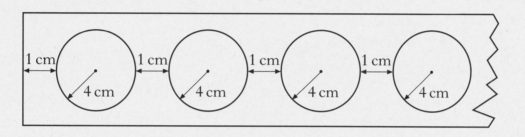

The radius of each top is 4 centimetres.

The gap between each top is 1 centimetre.

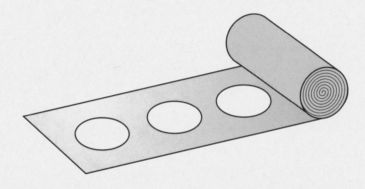

How many tops can be cut from a strip of foil 7 metres long?

4

Marks | KU | RE

12. A boat elevator is used to take a boat from the lower canal to the upper canal.

The boat elevator is in the shape of a triangle.

The length of the hypotenuse is 109 metres.

The height of the triangle is 45 metres.

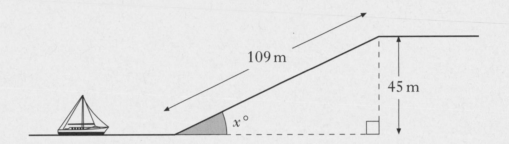

Calculate the size of the shaded angle $x°$.

3

Marks   KU   RE

13. A wheelie bin is in the shape of a cuboid.

The dimensions of the bin are:

- length 70 centimetres
- breadth 60 centimetres
- height 95 centimetres.

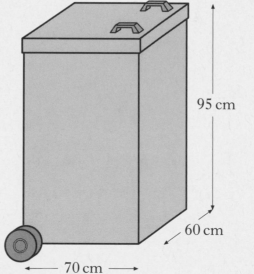

95 cm

60 cm

← 70 cm →

(*a*) Calculate the volume of the bin.

2

(*b*) The council is considering a new design of wheelie bin.

The new bin will have the same volume as the old one.

The base of the new bin is to be a square of side 55 centimetres.

Calculate the height of the new wheelie bin.

3

[END OF QUESTION PAPER]

ADDITIONAL SPACE FOR ANSWERS

2009

[BLANK PAGE]

FOR OFFICIAL USE

G

	KU	RE
Total marks		

2500/403

NATIONAL
QUALIFICATIONS
2009

WEDNESDAY, 6 MAY
10.40 AM – 11.15 AM

MATHEMATICS
STANDARD GRADE
General Level
Paper 1
Non-calculator

Fill in these boxes and read what is printed below.

Full name of centre

Town

Forename(s)

Surname

Date of birth
Day Month Year

Scottish candidate number

Number of seat

1 **You may not use a calculator.**

2 Answer as many questions as you can.

3 Write your working and answers in the spaces provided. Additional space is provided at the end of this question-answer book for use if required. If you use this space, write clearly the number of the question involved.

4 Full credit will be given only where the solution contains appropriate working.

5 Before leaving the examination room you must give this book to the invigilator. If you do not you may lose all the marks for this paper.

FORMULAE LIST

Circumference of a circle: $C = \pi d$

Area of a circle: $A = \pi r^2$

Curved surface area of a cylinder: $A = 2\pi rh$

Volume of a cylinder: $V = \pi r^2 h$

Volume of a triangular prism: $V = Ah$

Theorem of Pythagoras:

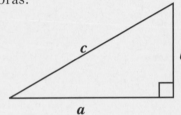

$$a^2 + b^2 = c^2$$

Trigonometric ratios
in a right angled
triangle:

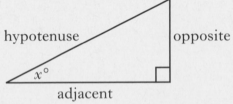

$$\tan x^\circ = \frac{\text{opposite}}{\text{adjacent}}$$

$$\sin x^\circ = \frac{\text{opposite}}{\text{hypotenuse}}$$

$$\cos x^\circ = \frac{\text{adjacent}}{\text{hypotenuse}}$$

Gradient:

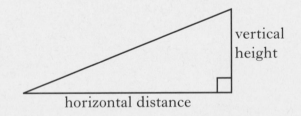

$$\text{Gradient} = \frac{\text{vertical height}}{\text{horizontal distance}}$$

DO NOT WRITE IN THIS MARGIN

Marks KU RE

1. Carry out the following calculations.

(*a*) 17·3 – 14·86 **1**

(*b*) 23 × 6000 **1**

(*c*) 256·9 ÷ 7 **1**

(*d*) 80% of 54 **2**

[Turn over

Marks | KU | RE

2. An old unit of measurement called a fluid ounce is equal to 0·0296 litres.

Write 0·0296 in scientific notation.

2

DO NOT
WRITE IN
THIS
MARGIN

Marks | KU | RE

3. Samira is designing a chain belt.

Each section of the belt is made from metal rings as shown below.

1 section, 4 rings

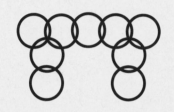

2 sections, 9 rings

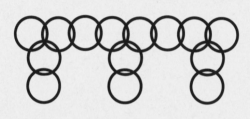

3 sections

(a) Complete the table below.

Number of sections (s)	1	2	3	4	5		11
Number of metal rings (r)	4	9					

2

(b) Write down a formula for calculating the number of rings (r), when you know the number of sections (s).

2

(c) Samira uses 79 rings to make her belt.

How many sections does her belt have?

2

4. A floor is to be tiled using tiles shaped like this.

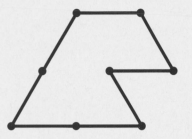

Here is part of the tiling.

Draw **four** more tiles to continue the tiling.

DO NOT
WRITE IN
THIS
MARGIN

Marks   KU   RI

3

Marks | KU | RE

5.   (*a*)   On the grid below, plot the points A(2, 6), B(8, 2) and C(6, –1).

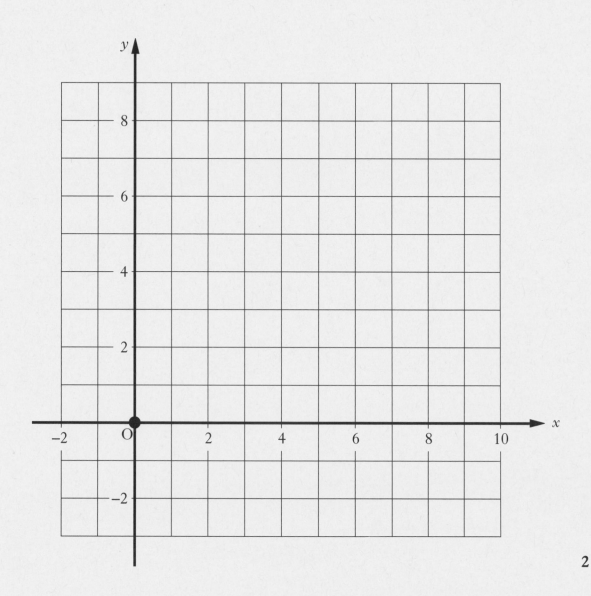

2

(*b*)   Plot a fourth point D so that ABCD is a rectangle.

1

(*c*)   On the grid, show the point where the diagonals of the rectangle intersect.

Write down the coordinates of this point.

2

Marks   KU   RE

6. In July the average temperature in Anchorage, Alaska is 9 °C.

By January the average temperature has fallen by 26 °C.

What is the average temperature in Anchorage in January?

2

     Page eight

Marks | KU | RE

7. Joe is making a fruit pudding on Scottish Master Chef.

In the fruit pudding recipe the ratio of raspberries to blackberries is 5:1.

Joe's fruit pudding must contain a **total** of 240 grams of fruit.

Calculate the weight of raspberries in his pudding.

3

[Turn over

DO NOT
WRITE IN
THIS
MARGIN

Marks KU | RE

8. Each pupil in a science class is growing a plant.

A few weeks later the height of each plant is measured.

The heights in centimetres are shown below.

6·3	5·4	5·8	7·0	6·2	7·6	8·3	8·4	5·3	8·8
8·5	5·6	6·8	6·5	6·1	6·7	7·4	7·6	5·3	

(*a*) Display these results in an ordered stem and leaf diagram.

3

(*b*) Find the median height.

1

DO NOT
WRITE IN
THIS
MARGIN

Marks | KU | RE

9. In the diagram below:

- triangle ABD is isosceles with AB = AD
- angle DAB = 34°
- angle ABC = 90°
- angle BCD = 20°.

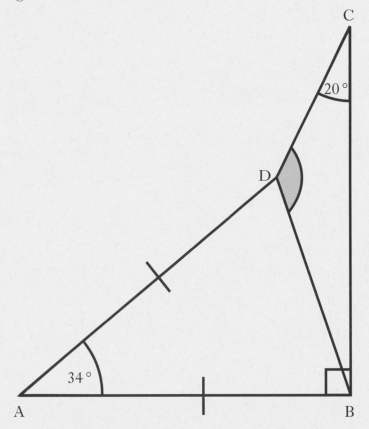

Calculate the size of the shaded angle BDC.

3

[*END OF QUESTION PAPER*]

ADDITIONAL SPACE FOR ANSWERS

FOR OFFICIAL USE

G

Total marks

KU	RE

2500/404

NATIONAL
QUALIFICATIONS
2009

WEDNESDAY, 6 MAY
11.35 AM – 12.30 PM

MATHEMATICS
STANDARD GRADE
General Level
Paper 2

Fill in these boxes and read what is printed below.

Full name of centre

Town

Forename(s)

Surname

Date of birth

Day Month Year

Scottish candidate number

Number of seat

1 **You may use a calculator.**

2 Answer as many questions as you can.

3 Write your working and answers in the spaces provided. Additional space is provided at the end of this question-answer book for use if required. If you use this space, write clearly the number of the question involved.

4 Full credit will be given only where the solution contains appropriate working.

5 Before leaving the examination room you must give this book to the invigilator. If you do not you may lose all the marks for this paper.

FORMULAE LIST

Circumference of a circle: $\qquad$ $C = \pi d$

Area of a circle: $\qquad$ $A = \pi r^2$

Curved surface area of a cylinder: $\qquad$ $A = 2\pi rh$

Volume of a cylinder: $\qquad$ $V = \pi r^2 h$

Volume of a triangular prism: $\qquad$ $V = Ah$

Theorem of Pythagoras:

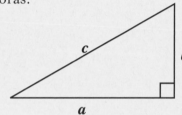

$$a^2 + b^2 = c^2$$

Trigonometric ratios in a right angled triangle:

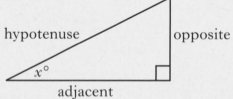

$$\tan x^\circ = \frac{\text{opposite}}{\text{adjacent}}$$

$$\sin x^\circ = \frac{\text{opposite}}{\text{hypotenuse}}$$

$$\cos x^\circ = \frac{\text{adjacent}}{\text{hypotenuse}}$$

Gradient:

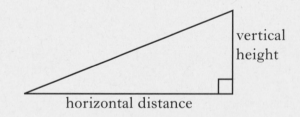

$$\textbf{Gradient} = \frac{\textbf{vertical height}}{\textbf{horizontal distance}}$$

Marks | KU | RE

1. Naveen drives from Dumfries to Manchester.

A 28 mile part of his journey is affected by roadworks.

It takes him 40 minutes to drive this part of his journey.

Calculate his average speed for this part of his journey.

Give your answer in miles per hour.

3

[Turn over

Marks KU RE

2. Helen travels between Glasgow and Edinburgh by train.

   She buys a monthly TravelPass which costs £264·30.

   A daily return ticket would cost £16·90.

   Last month Helen made 19 return journeys.

   How much did she save by buying the TravelPass?

3

Marks

3. A semi-circular window in the school assembly hall is made from three identical panes of glass.

During a recent storm one pane of glass was damaged.

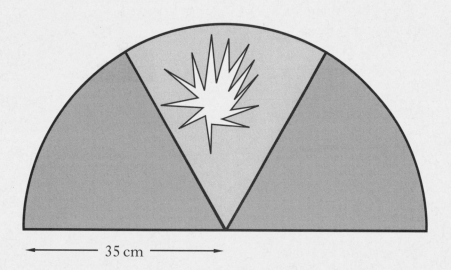

← 35 cm →

The semi-circle has a radius of 35 centimetres.

Calculate the area of the damaged pane of glass.

3

[**Turn over**

4. John is going to see a movie.

The movie has an evening and a late night showing.

	Evening showing	Late night showing
Start time	1750	
Finish time	2005	0110

(*a*) How long does the movie last?

1

(*b*) When does the late night showing start?

2

DO NOT WRITE IN THIS MARGIN

Marks KU RE

5. (*a*) Factorise

$$6c - 15d.$$

2

(*b*) Simplify

$$5(a + 1) + 2(5 - 2a).$$

3

[Turn over

DO NOT
WRITE IN
THIS
MARGIN

Marks   KU   RE

6. David is trying to decide which channel mixes to buy for his TV system.

The cost of each is:

- Drama Mix   £7
- Sport Mix   £20
- Movies Mix   £15
- Kids Mix   £12
- Music Mix   £10

He has decided to buy four different mixes.

One possible selection and its cost are shown in the table below.

(*a*) Complete the table showing all the possible selections and the cost of each.

Selections				Cost
Drama	Sport	Movies	Music	£52

3

(*b*) David can spend up to £55 for his selection.

Which selection can he **not** buy?

1

DO NOT WRITE IN THIS MARGIN

Marks KU RE

7. Last week Theresa asked 76 students to record how many hours they spent doing homework.

The results are shown below.

Homework hours	*Frequency*	*Homework hours × frequency*
1	16	
2	12	
3	18	
4	11	
5	8	
6	6	
7	5	
	Total = 76	Total =

Complete the above table and find the **mean** time spent on homework last week.

Round your answer to 1 decimal place.

4

Marks | KU | RE

8. A steel plate in the shape of an isosceles triangle is used to strengthen a bridge.

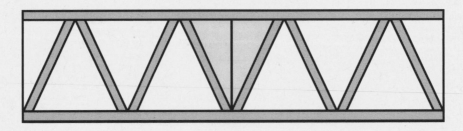

The dimensions of the isosceles triangle are shown below.

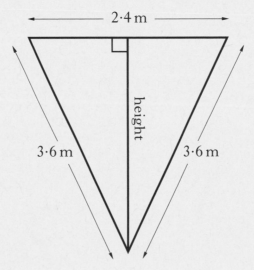

2·4 m

height

3·6 m   3·6 m

Calculate the height of the steel plate.

Do not use a scale drawing.

4

9.

Pizza Perfection — free delivery				
	Deep Base		Thin Base	
	9-inch	12-inch	9-inch	12-inch
Margherita	£3·60	£5·00	£3·30	£4·60
Mushroom	£4·25	£5·80	£4·15	£5·50
Pepperoni	£5·00	£6·30	£4·90	£6·00
Vegetarian	£5·05	£6·35	£4·95	£6·05
Hot Spicy	£5·15	£6·45	£5·05	£6·15

Iona and her friends order some pizzas to be delivered.

They order a 9-inch Hot Spicy deep base, a 12-inch Margherita deep base and two 12-inch Vegetarian thin base.

Find the total cost of the order.

3

[Turn over

Marks | KU | RE

10. Susan has £6200 in her Clydeside Bank account.

Clydeside Bank pays interest at 2·5% per annum.

Highland Bank pays interest at 3·7% per annum.

How much more money would Susan get in interest if she moved her £6200 to the Highland Bank for one year?

3

11. The shaded part of a garden light is triangular.

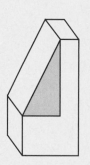

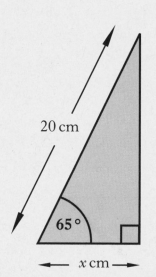

20 cm

65°

← *x* cm →

- the triangle is right angled
- the sloping edge is 20 centimetres long
- the angle between the base and the sloping edge is 65°.

Calculate the value of *x*.

3

Marks KU RE

12. The local council is installing a new children's playpark using a rubberised material.

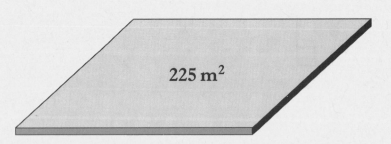

225 m²

The area of the rectangular playpark is 225 square metres.

The new playpark must have a depth of 12 centimetres.

The council has ordered 30 cubic metres of the rubberised material for the playpark.

Will this be enough?

Give a reason for your answer.

3

Marks   KU   RE

13. An off shore wind farm is on a bearing of 115° and at a distance of 90 kilometres from Eyemouth.

Using a scale of 1 centimetre to represent 10 kilometres, show the position of the wind farm on the diagram below.

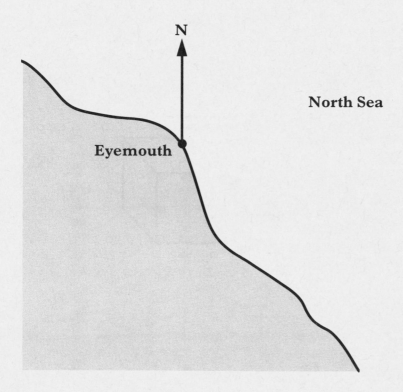

3

[Turn over for Question 14 on *Page sixteen*

Marks   KU   RE

14. The diagram below shows the net of a cube.

The total surface area of the cube is 150 square centimetres.

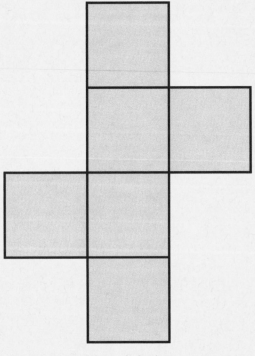

Net of Cube

Calculate the length of the side of the cube.

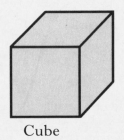

Cube

3

[END OF QUESTION PAPER]

ADDITIONAL SPACE FOR ANSWERS

ADDITIONAL SPACE FOR ANSWERS

ADDITIONAL SPACE FOR ANSWERS

[BLANK PAGE]

[BLANK PAGE]

FOR OFFICIAL USE

G

	KU	RE
Total marks		

2500/403

NATIONAL
QUALIFICATIONS
2010

WEDNESDAY, 5 MAY
10.40 AM – 11.15 AM

MATHEMATICS
STANDARD GRADE
General Level
Paper 1
Non-calculator

Fill in these boxes and read what is printed below.

Full name of centre

Town

Forename(s)

Surname

Date of birth

Day   Month   Year   Scottish candidate number   Number of seat

1. **You may not use a calculator.**

2. Answer as many questions as you can.

3. Write your working and answers in the spaces provided. Additional space is provided at the end of this question-answer book for use if required. If you use this space, write clearly the number of the question involved.

4. Full credit will be given only where the solution contains appropriate working.

5. Before leaving the examination room you must give this book to the Invigilator. If you do not, you may lose all the marks for this paper.

FORMULAE LIST

Circumference of a circle: $C = \pi d$

Area of a circle: $A = \pi r^2$

Curved surface area of a cylinder: $A = 2\pi rh$

Volume of a cylinder: $V = \pi r^2 h$

Volume of a triangular prism: $V = Ah$

Theorem of Pythagoras:

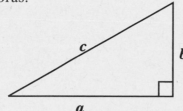

$$a^2 + b^2 = c^2$$

Trigonometric ratios
in a right angled
triangle:

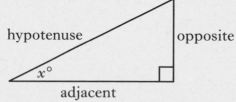

$$\tan x° = \frac{\text{opposite}}{\text{adjacent}}$$

$$\sin x° = \frac{\text{opposite}}{\text{hypotenuse}}$$

$$\cos x° = \frac{\text{adjacent}}{\text{hypotenuse}}$$

Gradient:

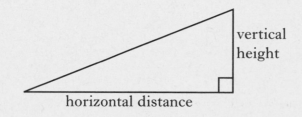

$$\text{Gradient} = \frac{\text{vertical height}}{\text{horizontal distance}}$$

DO NOT WRITE IN THIS MARGIN

Marks | KU | RE

1. Carry out the following calculations.

(*a*) $9{\cdot}32 - 5{\cdot}6 + 4{\cdot}27$ **1**

(*b*) $37{\cdot}6 \times 8$ **1**

(*c*) $2680 \div 400$ **1**

(*d*) $7 \times 2\frac{1}{3}$ **2**

[Turn over

Marks | KU | RE

2. The space shuttle programme costs $5800 million.

Write this number in scientific notation.

2

3. One day last February, Anna compared the temperature in Edinburgh with the temperature in Montreal.

The temperature in Edinburgh was 8 °C.

The temperature in Montreal was –15 °C.

Find the difference between these temperatures.

2

4. Complete this design so that the dotted line is an axis of symmetry.

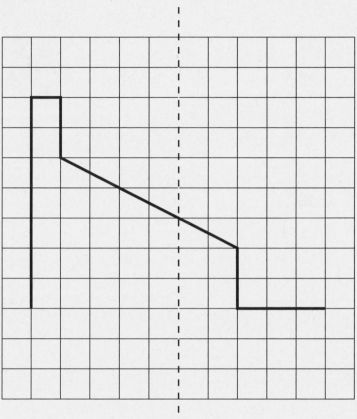

3

[Turn over

Marks

KU | RE

5. Karen asked her class to note the number of songs they downloaded to their phones in the last month.

The answers are shown below.

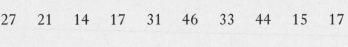

| 14 | 16 | 15 | 26 | 11 | 32 | 12 | 13 | 42 | 51 |
| 27 | 21 | 14 | 17 | 31 | 46 | 33 | 44 | 15 | 17 |

Display these answers in an ordered stem and leaf diagram.

3

DO NOT
WRITE IN
THIS
MARGIN

Marks | KU | RE

6. Carla is laying a path in a nursery school.

She is using a mixture of alphabet tiles and coloured tiles.

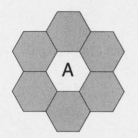

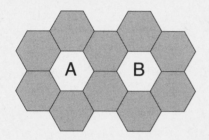

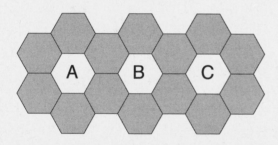

(a) Complete the table below.

Number of alphabet tiles (a)	1	2	3	4	5		12
Number of coloured tiles (c)	6	10					

2

(b) Write down a formula for calculating the number of coloured tiles (c) when you know the number of alphabet tiles (a).

2

(c) Carla uses 86 coloured tiles to make the path.

How many alphabet tiles will be in the path?

2

Marks | KU | RE

7. When on holiday in Spain, Sandy sees a pair of jeans priced at 65 euros.

   Sandy knows that he gets 13 euros for £10.

   What is the price of the jeans in pounds?

65 euros

3

8. The price of a laptop is reduced from £400 to £320.

   Calculate the percentage reduction in the price of the laptop.

£400
£320

3

9. The diagram shows a triangular prism.

The dimensions are given on the diagram.

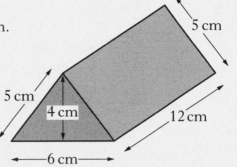

A **net** of this triangular prism is shown below.

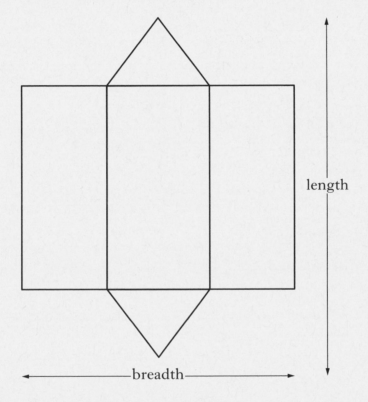

Calculate the length and breadth of this net.

2

[Turn over for Question 10 on *Page ten*

DO NOT
WRITE IN
THIS
MARGIN

Marks   KU   RE

10.

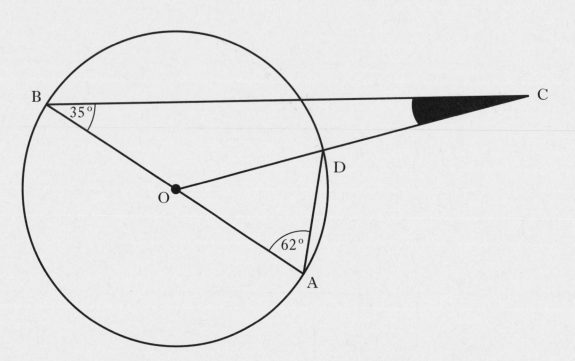

In the diagram above:

- AB is a diameter of the circle with centre O

- OC intersects the circle at D

- Angle ABC = 35°

- Angle BAD = 62°

Calculate the size of the shaded angle.

3

[END OF QUESTION PAPER]

ADDITIONAL SPACE FOR ANSWERS

[BLANK PAGE]

FOR OFFICIAL USE

G

	KU	RE
Total marks		

2500/404

NATIONAL
QUALIFICATIONS
2010

WEDNESDAY, 5 MAY
11.35 AM – 12.30 PM

MATHEMATICS
STANDARD GRADE
General Level
Paper 2

Fill in these boxes and read what is printed below.

Full name of centre

Town

Forename(s)

Surname

Date of birth

Day	Month	Year	Scottish candidate number	Number of seat

1. **You may use a calculator.**

2. Answer as many questions as you can.

3. Write your working and answers in the spaces provided. Additional space is provided at the end of this question-answer book for use if required. If you use this space, write clearly the number of the question involved.

4. Full credit will be given only where the solution contains appropriate working.

5. Before leaving the examination room you must give this book to the Invigilator. If you do not, you may lose all the marks for this paper.

FORMULAE LIST

Circumference of a circle: $C = \pi d$

Area of a circle: $A = \pi r^2$

Curved surface area of a cylinder: $A = 2\pi rh$

Volume of a cylinder: $V = \pi r^2 h$

Volume of a triangular prism: $V = Ah$

Theorem of Pythagoras:

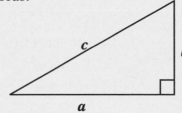

$$a^2 + b^2 = c^2$$

Trigonometric ratios
in a right angled
triangle:

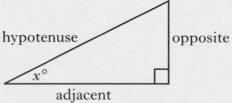

$$\tan x^\circ = \frac{\text{opposite}}{\text{adjacent}}$$

$$\sin x^\circ = \frac{\text{opposite}}{\text{hypotenuse}}$$

$$\cos x^\circ = \frac{\text{adjacent}}{\text{hypotenuse}}$$

Gradient:

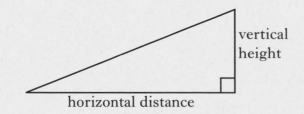

$$\textbf{Gradient} = \frac{\textbf{vertical height}}{\textbf{horizontal distance}}$$

Marks

KU	RE

1. Ten people were asked to guess the number of coffee beans in a jar.

   Their guesses were:

   310   260   198   250   275   300   245   225   310   200

   (*a*)   What is the range of this data?

                                                      1

   (*b*)   Find the median.

                                                      2

[Turn over

DO NOT
WRITE IN
THIS
MARGIN

Marks   KU   RE

2. Mr and Mrs Kapela book a cruise to Bruges for themselves and their three children.

- They depart on 27 June

  Mr and Mrs Kapela share an outside cabin and their three children share an inside cabin

  There is a 20% discount for each child

Calculate the total cost of the cruise.

Mini Cruise to Bruges, Belgium		
	Price per person	
Departure Date	Inside Cabin (£)	Outside Cabin (£)
16 May	236	250
30 May	244	274
13 June	266	300
27 June	275	310
12 July	291	325
26 July	312	355
9 Aug	327	370

3

Marks | KU | RE

3. As part of his healthy diet, Tomas has decided to buy fruit in his weekly shopping.

His favourite fruits and their costs per pack are given in the table below.

Fruit	Cost
Apples	£1·25
Oranges	£1·20
Grapes	£2·49
Pears	£1·56
Melon	£1·98

He wants to

- buy 3 different packs of fruit
- spend a maximum of £5 on fruit.

One possible selection and its cost are shown in the table below.

Complete the table to show all of Tomas's possible selections and their cost.

Apples	Oranges	Grapes	Pears	Melon	Cost
✓	✓		✓		£4·01

4

[Turn over

DO NOT
WRITE IN
THIS
MARGIN

Marks   KU   RE

4. (*a*) Complete the table below for $y = 2x - 3$.

x	−1	1	3
y			

2

(*b*) Using the table in part (*a*), draw the graph of the line $y = 2x - 3$ on the grid below.

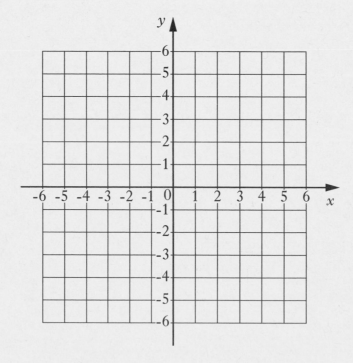

2

Marks | KU | RE

5. For safety reasons the speed limit outside Fairfield Park is 20 miles per hour.

The distance between the speed limit signs outside Fairfield Park is half a mile.

A van took 2 minutes to travel between these signs.

Was the van travelling at a safe speed?

Give a reason for your answer.

3

[Turn over

DO NOT WRITE IN THIS MARGIN

Marks KU RE

6. (*a*) Simplify

$$8(c-3)+5(c+2).$$

3

(*b*) Solve algebraically

$$25 = 7x + 4.$$

2

Marks | KU | RE

7. Rowan wants to buy 13 theatre tickets.

   The price of one ticket is £12·50.

   The theatre has a special online offer of four tickets for the price of three.

   Rowan makes use of the special online offer.

   How much does Rowan pay for the 13 theatre tickets?

Online Ticket Offer
4 for the price of 3

3

[Turn over

DO NOT
WRITE IN
THIS
MARGIN

Marks | KU | RE

8. A survey of 1800 first time voters was carried out.

The pie chart below shows how they would vote at the next election.

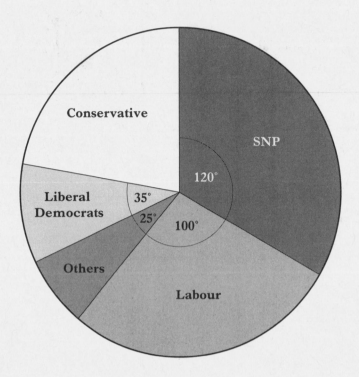

How many of the 1800 first time voters would vote Conservative?

3

DO NOT
WRITE IN
THIS
MARGIN

Marks   KU   RE

9. A tennis court is 11 metres wide.

It has an area of 264 square metres.

11 m

Calculate the perimeter of the tennis court.

3

[Turn over

Marks | KU | RE

10. Ahmed is making a frame to strengthen a stairway in a shopping centre.

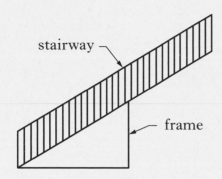

He needs to know the angle the stairway makes with the floor, as shown in the diagram below.

The hypotenuse of the frame is 5·2 m and the horizontal distance is 4·5 m.

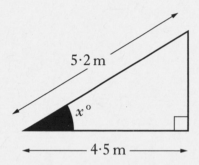

Calculate the size of the shaded angle $x°$.

3

Marks

KU | RE

11. A climber needs to be rescued.

His position from the helicopter base is marked on the map.

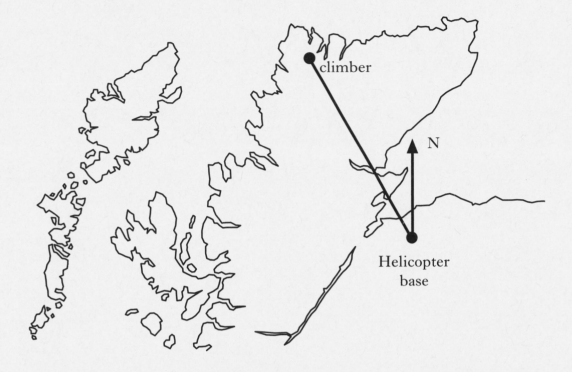

(*a*) Using a scale of 1 centimetre to 15 kilometres, calculate the distance of the climber from the helicopter base.

1

(*b*) Find the bearing of the climber from the helicopter base.

2

[Turn over

Marks   KU   RE

12. An earring in the shape of an isosceles triangle is made from silver wire.

The dimensions of the earring are shown on the diagram below.

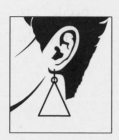

5 cm

3 cm

Calculate the length of silver wire needed to make a **pair** of earrings.

Do not use a scale drawing.

4

Marks | KU | RE

13. A plastic speed bump in the shape of a half cylinder is used to slow traffic outside a Primary School.

The speed bump has radius of 10 centimetres and a length of 7 metres as shown in the diagram below.

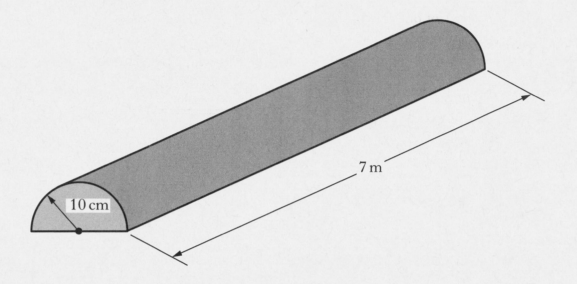

7 m

10 cm

Calculate the volume of plastic used to make the speed bump.

3

[Turn over for Question 14 on *Page sixteen*

Marks | KU | RE

14. Liam buys a new stereo using the monthly payment plan.

The cash price of the stereo is £360.

The total cost of the monthly payment plan is **5% more than the cash price**.

Liam pays a deposit of one fifth of the cash price followed by 30 equal monthly payments.

Cash Price £360

Monthly Payment Plan
Deposit ⅕ of cash price
and 30 monthly payments

How much will Liam pay each month?

4

[END OF QUESTION PAPER]

ADDITIONAL SPACE FOR ANSWERS

ADDITIONAL SPACE FOR ANSWERS

ADDITIONAL SPACE FOR ANSWERS

[BLANK PAGE]